D1186585

BRUTAL SIMPLICITY OF THOUGHT

This book started life as a training manual for Saatchi employees.
Its approach shaped the Saatchistory for 40 years. Its principles permeate the culture,
philosophy and structure of one of the best known corporate brands.

Contributors: *Asli Azizoglu • Oliver Batho • Natalie Blass • Curtis Brittles • Liam Burrell • Samantha Carter Bernard Chan • Nikki Cochrane • Mathew Conrad • John Cunningham • Bryony Dellow • Simon Dicketts Graham Fink • L Fok • Bill Gallacher • Mark Goodwin • Monique Henry • Stephanie Illingworth • Phil Jackson Rachel Joice • Camilla Knight • Imogen Landy • Terry Leppard • Choi Liu • Peter Liu • Moray MacLennan Gary Monaghan • Jonathan Muddell • Andy Ng • Xavier Obon • Mark Petty • Bruce Pywell • Maurice Saatchi Mandy Saggu • Leanne Silman • Rupert Simonds-Gooding • Jeremy Sinclair • Darius Tang • Steve Tribe Simon Warden • Orlando Warner • Sophia Wetherell • Tim Wilks • Michael Wilton • Joke Wong • Dan Younger*

If you would like to contribute to any future edition please go to brutalsimplicityofthought.com
Keep it simple.

This book celebrates moments when Brutal Simplicity of Thought changed the world, and proved that nothing is impossible.

A few words on simplicity

If you want your work to achieve the impossible, you will need Brutal Simplicity of Thought.

You will need a deep distaste for waffle, vagueness, platitudes and flim flam – a strong preference to get to the point.

Your mind will become a threshing machine, sorting the intellectual wheat from the chaff.

Winston Churchill was a great believer in simplicity. He liked to quote Blaise Pascal's letter to a friend that started:

I didn't have time to write a short letter, so I wrote a long one instead

He knew that to achieve simplicity is very hard. He understood that it required what Bertrand Russell called:

The painful necessity of thought

Simplicity is more than a discipline: it is a test. It forces exactitude or it annihilates. It accelerates failure when a cause is weak, and it clarifies and strengthens a cause that is strong.

When President Roosevelt wanted to persuade a profoundly isolationist America to help Britain in her hour of need, he invented a simple phrase to help him do it. He called his policy:

Lend-Lease

And he used simple language to express it:

'Suppose my neighbour's home catches fire, and I have a length of garden hose … if he can take my garden hose and connect it up with his hydrant, I may help him to put out the fire … I don't say to him before that operation, "Neighbour, my garden hose cost me $15; you have to pay me $15 for it" … I don't want $15 – I want my garden hose back after the fire is over.'

That's how it was done. A simple story of a fire and a hose. The rest is history.

The most powerful rallying cries are simple and to the point:

Your country needs you!

No taxation without representation!

One man! One vote!

There was nothing complicated about:

Liberté, Egalité, Fraternité

Nobody had to explain what it meant when they heard John F. Kennedy say:

The torch has been passed to a new generation of Americans

Or when they read on the Statue of Liberty:

Give me your tired, your poor. Your huddled masses yearning to breathe free

Nobody needed further elucidation when they heard:

Do unto others as you would be done by

Or when Martin Luther King said:

I have a dream

In all aspects of life, simplicity rules. It means the only possible words in the only possible order.

Simplicity in poetry. John Keats was sitting in a coffee shop with his friend, Stephens.

He was writing. He said:

> *A thing of beauty is a constant joy. What think you of that, Stephens?*

No response from Stephens. Keats carried on. Half an hour later, Keats said:

> *A thing of beauty is a joy forever*

That, his friend said, will last forever. And it did.

Simplicity in art. Delacroix explained:

> *If you are not skilful enough to sketch a man falling out of the window during the time it takes to get him from the fifth story to the ground, then you will never be able to produce monumental work*

Simplicity in prose. Is it any wonder that Kafka lives forever, when you consider the opening words of *The Trial*?

> *Someone must have laid a false accusation against Joseph K because one morning he was arrested without having done anything wrong*

Simplicity in drama. Could Shakespeare have imagined that *Hamlet* would be the most performed play in history; that four hundred years later, there would be a performance of *Hamlet*, somewhere in the world, every minute, because he captured the human dilemma in ten words:

To be or not to be, that is the question

Simplicity in politics. During Britain's darkest hour, Winston Churchill was presented with the proposal for a Local Defence Volunteers Force, to be Britain's last stand in the event of a German invasion. The LDVF. He liked the plan. And approved it. But he didn't like the name. He changed it to:

The Home Guard

And so it became.

The post war 1918 general election was won by Lloyd George, with five words:

A land fit for heroes

In the post war 1945 General Election, Clement Atlee defeated the war hero Winston Churchill with nine words:

We won the war. And now - win the peace

The Conservatives were helped to win the 1979 General Election by three words:

Labour isn't working

You hear it said that this search for simplicity is insulting the intelligence of the public, or treating them like morons. On the contrary, it is a mark of respect for the listener. The world is always short of time, so a précis is a form of good manners.

At this point, you are thinking what has all this to do with the time of day? Words spell money.

Simplicity in business. Every day, a blind man sat on the pavement in Central Park. He had his hat in front of him, begging for money. A sign read:

I am blind

Passers-by ignored him. One day, an advertising man saw his plight. He altered the wording on his sign and the cash started pouring into the hat. What had he done?

He had changed the sign to read:

It is spring and I am blind

When William Procter and James Gamble started Procter and Gamble, they only had one insignificant product – Ivory Bar Soap. Until they added the slogan:

$99^{44}/_{100}$% *Pure*

That was the beginning of the P&G legend.

Simplicity rules. Consider the three iconic documents in Western civilisation. There really are only three of them. And they really did change the world. Their aim was revolution. Their effect was revelation. You need only look at them to be inspired. You will be deeply affected by all three. To read them afresh is to understand the power of simplicity. You don't need a Harvard Ph.D to follow any of them.

Their opening and closing words say it all. They are of course:

The Sermon on the Mount, by Jesus Christ

The Declaration of Independence, by the Founding Fathers of America.

The Communist Manifesto, by Karl Marx and Friedrich Engels

The first founded perhaps the greatest religion ever seen:

Open: And seeing the multitudes … he opened his mouth and taught them, saying …

Close: And it came to pass ... people were astonished at his doctrine

The second made one country into a superpower:

Open: We, the people, hold these truths to be self-evident

Close: …we mutually pledge to each other our lives, our fortunes and our sacred honour

And the third launched what Isaiah Berlin called the greatest organised social movement of all time, greater perhaps than the rise of Christianity against paganism:

Open: The history of all previously existing society is the history of class struggle

Close: Workers of the world unite! You have nothing to lose but your chains

Nobody can resist that kind of simplicity. Its reach is global. It strikes a chord in humans everywhere. Nobody is immune.

Their Brutal Simplicity of Thought allowed them to change the world.

With Brutal Simplicity of Thought, nothing is impossible.

Happy ending. There is an unexpected by-product of this process; it makes people happy. It enables the human mind to function at its best, and to be supremely effective.

It allows you to have a romantic belief in your ability to change the world by an act of breathtakingly brutal simplicity. It is a licence to reject the status quo. It leads to a determined conviction that you, acting alone or almost single-handedly, can make what seems highly improbable, in fact happen.

So that even the meekest can meet life with the possibility of mastering its difficulties.

The people who do not have such belief are miserable. They are mere men of commerce: non-believers, empty suits.

By contrast, such men and women as you, find happiness in transforming one form of life into another. You know you can permanently and radically alter the outlook and values of a significant body of human beings.

You will have power through what John F. Kennedy called:

The mastery of the inside of men's minds

Particularly your own…

Maurice Saatchi
September 2011

First draft notes for Maurice Saatchi's speech
at the Saatchistory 40th anniversary party
on October 9th 2010

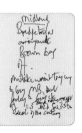

Final version

How do you hear from the dead?

How do you speak to the unborn?

Without writing, every generation would have to start from scratch. Philosophers would have no Plato; mathematicians, no Newton; scientists, no Einstein; actors, no Shakespeare. And your descendents wouldn't even know where you were buried. Or that you lived at all. But more important, we wouldn't be meeting on this page now.

How do you carry a cow in your pocket?

What if what you make, isn't wanted by the person whose goods you want? Or what if you didn't want, what the person who wanted your goods made? You need money. Then anyone can trade with ease with anyone - who has money. Example: for £10.00, rather than 22.72 pints of milk, this handsome book can be yours. Money is mother and father of specialisation and therefore, all progress.

LIBERTÉ,
ÉGALITÉ,
FRATERNITÉ

How did three words kill the French aristocracy?

The French Revolution shook the political and intellectual landscape of Europe. And it was behind these three words that the revolution took shape. They captured the imagination of the working class, and drove the revolution towards the fall of the Bourbon monarchy, the execution of Louis XVI and eight months later, the Reign of Terror.

How did two wheels emancipate women?

The first bicycle appeared in 1817, when Baron Karl von Drais presented his 'draisine' or 'walking machine.' This lead to the 'safety bicycle' in the mid 1880s, giving women's activists in the US and UK in particular, a means of travel that matched their ambitions. No longer were they beholden to fathers, brothers or husbands to get around. The bicycle became their freedom machine.

How do you catalogue everything?

N. Joseph Woodland took his inspiration from Morse code. He formed the first barcode out of sand on the beach when he 'just extended the dots and dashes downwards and made narrow lines and wide lines out of them'. In so doing, he created a unique, universal and cheap method of cataloguing any product anywhere. See back cover for a further riveting example.

How do you get a country to work an hour earlier?

Announce that the clocks are going to change at a certain time. In 1895, George Vernon Hudson regretted the number of wasted daylight hours in summer. He proposed Daylight Saving Time. By advancing clocks in summer, at a stroke, there would be more light available and therefore more work done, using less energy.

How do you draw like a child?

'It took me a whole lifetime to learn how to draw like a child.' Picasso is celebrated as one of the most respected artists of all time. Central to his legacy was his determination not to follow the artistic provenance of Raphael, but instead to reduce everything down to a child's scribble. This he saw as the greatest challenge. Picasso demonstrated the beauty of complete essentialism. Reduction to its core parts, where art becomes nothing but line and form.

How did an Irish Pointer discover Velcro?

The idea for Velcro, conceived by George de Mestral, occurred whilst out walking his dog. Burdock seeds were always getting caught in his dog's fur as she ran through the fields. De Mestral, an engineer, inspected further, and found the seeds hooked onto the fur with a series of microscopic loops. And so Velcro was born. It would go on to be used extensively, from children's trainers to boots for the moon landing. Good dog.

How did one good turn make the paper clip?

There have been myriads of designs of paper clip, but one reigns supreme: the Gem, created by William Middlebrook. Earlier, American Samuel B. Fay had been granted the first patent for a 'ticket fastener', subsequently marketed as the 'Cinch' paper clip. It was, however, flawed because of one simple omission: the last turn of the wire. That, in turn, wrote the Gem into history.

How do you make music
with a symbol and a line?

The brutal brilliance of musical notation lies in its ability to visualise all music within a single system. Every nuance, all emotion and beauty, is recorded and poised to be played, using no more than a symbol and a line. Whether it's grime, blues, rock, classical or techno – the same system of notation can be used to capture and play any tune time and time again.

Can it be quicker to go through solid rock?

The congestion of London had plagued travellers for hundreds of years. The network of tightly packed streets made movement across the capital hugely problematic. Charles Pearson didn't create wider roads to travel, or one-way traffic, or roundabouts, or bus lanes. Pearson's idea was that people would get to their destination more quickly and directly by travelling underground – 'trains in drains'. Last year, London Underground carried over a billion people under London ground.

How do you wage war without violence?

Martin Luther King's dream was an equal United States of America. Peace held the key to this dream – not because peace was the end goal, but because it was peace that would be a 'means by which we arrive at that goal'. When it came to making his dream come true, peace was his weapon of choice - as it had been Gandhi's before him.

How did cats' eyes replace tramlines?

Early motorists used the reflections of their cars' lights in tramlines to guide them home in the dark, but when the tramlines were removed in the early 1930s, they were lost. One night, Percy Shaw saw his headlamps reflected in a cat's eyes, inspiring him to invent his 'self-wiping road studs'.

—•—• —— ——— • —• •••— •• •— •—• ••— —•

—— • —• ••• •— •— •—— • ••• •• —•

•—• • —• •—• • •—•• •— —••• •—• •— ••• ••—••

How can you send a message without words?

Until the 1800s messages were delivered by hand or pigeon. The advent of electricity made it possible to send an electric current at great speed, but the technology of the time couldn't print characters in readable form. In 1836 Morse and Vail solved the problem by developing an alphabet based on rhythm. They sent pulses of electrical current to an electromagnet that was located at the receiving end of the telegraph wire. Morse code, the language without words, had been born.

0

How do you turn one into ten by adding nothing?

If you were asked to invent a system for writing down numbers, you might do 1 mark for one, 2 for two and so on. But would you have a symbol for no number, for nothing? The genius of 0 is that it holds the place of empty values – so for example in 100 we know that there are no values in the tens or ones categories. We learnt it from the Arabs and so called the system Arabic, they learnt it from the Indians and so called it Indian. Nothing is more brilliant.

How do you make densely populated areas pleasant?

The third millennium BC that was dubbed the 'age of cleanliness', and for good reason. The toilets of Mohenjo-Daro in what is modern day Pakistan were some of the most advanced lavatories ever built. These toilets were incorporated into the outer walls of houses, and wouldn't be out of place today. They were constructed with bricks, linked to drainage systems and were even covered with wooden seats. The toilet, and the domestication of modern civilization, had arrived.

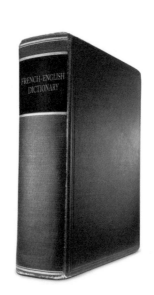

Comment parlez-vous aux étrangers?

What did we do before the dictionary? How did we communicate with a foreigner? The dictionary opened up communication on an international level. Until then, communication was locked in pockets of regional dialects. We now had the foundations for international dialogue – a tool to contrast and learn new languages. The oldest known bilingual dictionary was discovered in Syria, and dates back to 2300BC.

How do you have tea without leaves?

Struggling to make ends meet, Thomas Sullivan, a New York coffee merchant turned to tea, and sent out samples in small silk sachets, rather than as loose tea in expensive tins. His penny-pinching was misunderstood by his customers who failed to realise that they were supposed to cut open the sachet and empty the contents into a pot before brewing their tea. The teabag was born.

How do you get the customer to make your products?

The manufacturer makes the product, the consumer buys it. That was the philosophy of commerce. Or at least it was until the mid 1950s, when Gillis Lundgren was trying to pack a new table into the back of his car. It didn't fit, so he took the legs off, with the idea of reassembling it later. Realising he was on to something, he took the idea to his employers at Ikea. It was 1956, and 'flat pack' philosophy had taken hold.

HOW DO YOU MAKE MONEY
FROM PEOPLE'S MITSAKES?

In 1770, an English engineer called Edward Nairne discovered rubber's erasing properties when he accidentally picked up a piece of rubber instead of breadcrumbs, which had been traditionally used to remove small errors on parchment or papyrus documents. It was the first practical application of the substance in Europe, and coincidentally the 'rubbing' movement gave the rubber its English name.

What laid the foundations for the skyscraper?

It wasn't until the 1850's that Elisha Otis unlocked the true potential of the sky. Worried about the dangerous nature of lifts, yet needing to move heavy machinery and workers to the top floor of his factory, he sought to create a 'safety elevator'. His designs specified a teeth-related braking device that would clamp onto the rails of the elevator should the supporting cables break. His breakthrough allowed mankind to rise upwards.

How do you hold a building together with thread?

The Greek polymath Archimedes is credited with the invention of the water screw. By the first century BC screws were widely used throughout the Mediterranean in wine and oil presses. They would go on to become universal in application. Standard thread would be established in every country and industry in the world – where just the right angle of thread on a screw would make for easier production and stronger fastening.

Which animal's bladder unites and divides the world?

With an estimated 3.5 billion fans, football is the most popular sport on the planet. As early as the second century BC, a game akin to football was played by the Han Dynasty, where they used to kick a ball through a hole in a piece of silk cloth suspended nine metres above the ground. Similarly, in England in the Middle Ages a pig's bladder was inflated for 'mob football', involving an unlimited number of players from opposing villages. The modern game of football was born.

www.hat's the quickest way to find anything?

You know you've made it when your name becomes a verb. To Hoover is to clean. To Xerox is to copy. To Google is to find.

How do you persuade customers to buy more than they can carry?

In 1912, a small grocery store owner noticed that his customers were only buying as many products as they could carry. His solution was a neat, inexpensive paper bag with a cord running through it for strength. Walter H. Deubner had invented the carrier bag, and changed the way the world shopped.

IN CONGRESS, JULY 4, 1776.

The unanimous Declaration of the thirteen united States of America,

How do you create the world's most powerful nation?

On a single sheet of paper, the most powerful nation in the history of the world was born. Fifty-six men signed the American Dream into scripture. They declared that 'life, liberty and the pursuit of happiness' are the inalienable rights – and that all men are created equal. It was the spark that would drive millions of people to follow their dreams.

In 1851, Elias Howe patented an 'automatic, continuous clothing closure', that was very similar to the mechanism we know today as the zip. It wouldn't be until the 1930s that the zipper would be embraced commercially by the fashion industry – primarily as a method of getting kids to dress themselves. However in 1937 the zipper beat the button in the annual 'battle of the fly', and as *Esquire* said, one of its primary virtues was to exclude the 'possibility of unintentional and embarrassing disarray.'

How can you make night day?

The earliest forms of candle were made from whale fat, and originated in China around 200 BC. Usage spread from mere illumination, to the accurate telling of time – through the uniform burning of the candle. Beyond this, the candle took on hugely important symbolic resonance in religion and science, the single light coming to represent the deity and the spiritual within the darkness. Not bad for mammal fat.

How can colour control human behaviour?

We now know there is a psychology to colours. They can elicit particular responses in us that we have little control over. For example, red light is proven to increase the heartbeat. The first traffic light was established in 1868 outside the Houses of Parliament. Simple revolving gas lights, one red and one green. Unfortunately on 2nd January 1869 it exploded, killing the attendant policeman.

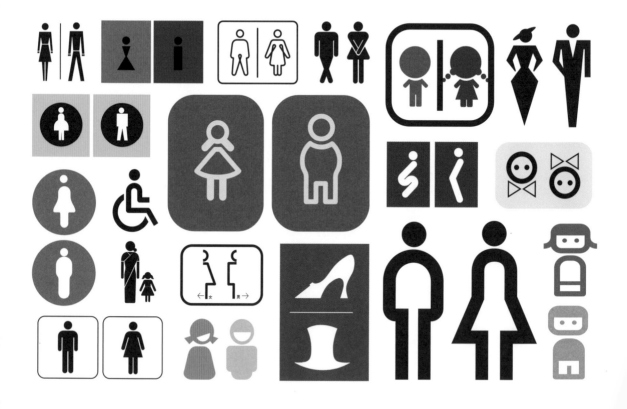

How do you go to the toilet anywhere?

The school of semiotics is the study of signs and cultural communication systems. The importance of signs was noted as early as the Greeks. Aristotle and Plato discussed the relationship between signs and the world. The best signs transcend traditional teaching, and elicit an immediate understanding from the viewer.

Why me?

If you believe the fates are against you, you are wrong. Brutal Simplicity of Thought can make a mortal, immortal; a slave, a master; and a nobody into somebody. It is the key to success and you are lucky, you have it in your hands.

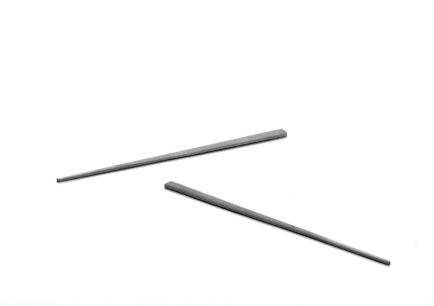

How do two sticks feed a continent?

It is believed chopsticks were developed 3,000–4,000 years ago. Early Asian man would retrieve his food from the fire with sticks or branches. The popularity of Confucianism is credited with spreading chopsticks around Asia as the primary eating utensils. Confucius taught, 'The honourable and upright man keeps well away from the slaughter house and the kitchen. And he allows no knives on his table.' He was vegetarian.

How can a caterpillar change the world?

According to Chinese mythology, the 27th century BC marked the birth of sericulture, the rearing of silk worms for raw silk production. The story has it, that a silk cocoon fell into Empress Leizu's tea, and as she unravelled it from the cup, she had the idea to weave it. Silk and the Silk Road became a focus of commerce for hundreds of years. In time, silk would conquer the world as people were seduced by its sensuous look and feel.

How do you measure volume?

Archimedes had his epiphany in the bath. He had been asked to find out if a local goldsmith had substituted non-gold elements into a crown for the monarch. Thinking on the problem in the bath he had his 'Eureka' moment. His own volume displaced what seemed to be an equal volume of water. He surmised that dividing the mass of the crown by the volume of water it displaced would give him the density. He would later confirm that silver had indeed been added, and so the dishonest goldsmith was found out.

How do you make people worship an instrument of torture?

Sacrifice is a powerful force. The idea that someone would surrender his life for the sake of others. What could better capture the Christian ideal than the Son of Man demonstrating love for mankind over and above his own welfare? In giving up his life, the symbol of his demise would unify billions for thousands of years – a constant reminder that Christian love is not selfish love.

What holds the secret to time travel?

Prior to the invention of the mainspring in 15th century, clocks were powered by hanging weights which meant they had to be in a fixed mounted position. Time stood still and so, often varied from town to town. The use of a tightly wound spring meant clocks were no longer confined. From this moment onwards, time travel became possible.

How do you make
a plate edible?

The ice cream cone made its first appearance in the 19th century. According to one account, the Syrian Ernest Hamwi came to the aid of the neighbouring ice cream vendor who had run out of plates. He simply folded one of his waffles into a cone shape. Another school of thought claims vendors in Western Europe grew tired of washing glass plates that were licked clean. Cones provided a more efficient way of serving their *coupe de glace*.

Why limit ourselves to a third of the world?

From the early drawings of Leonardo, to the first commercially successful diving apparatus developed by Jacques Cousteau and Emile Gagnan, there has been an urge to explore the deep. Cousteau went on to extend the idea and reveal to the world a previously unknown and inaccessible wonderland. The depths of the ocean were unlocked by unlocking underwater breathing.

How can a photo change a high street?

The astronauts set off to discover the secrets of outer space. Yet it can be argued mankind learned more about man on earth than it did about the man in the moon. 'I saw the earth without the scars of national boundaries' wrote one pioneer. So began the earnest global conquest by Nike, McDonald's, Gap, Benetton, Starbucks, Saatchi, Zara, 'Les Mis', Sony, Apple …

How do you create instant boundaries?

In 1867, Lucien B Smith was granted the patent for barbed or barb wire. The invention instantly reduced the cost of enclosing land, ultimately leading to the Big Die Up of 1885, when the traditional migration of cattle from the Northern Plains of America to the warmer Southern Plains was blocked by barb wire fencing. Three quarters of all herds were decimated. Barbed wire would become synonymous with some of the most polarizing moments in human history. The fences of Auschwitz; the walls of Guantanamo Bay; dividing lines in Palestine. It has excelled at creating boundaries.

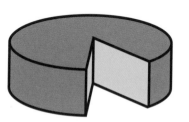

How can a pie save thousands of lives?

Florence Nightingale is revered as the mother of modern nursing – but she was also a pioneer in the visualisation of statistics. During the Crimean War she used the pie chart, a little known format developed by William Playfair, to demonstrate to politicians how thousands of soldiers were dying not on the front, but in hospital beds from preventable diseases. She proved that conditions in hospitals needed a radical overhaul. And the lives of thousands of soldiers were saved by simple hygiene and what now seems common sense.

How do you control an ostrich?

Ostriches are notoriously nervous. They are also capable of running at up to 72km per hour, and if cornered kicking viciously. Put a sock on their heads however and the ostrich becomes completely docile. It seems it believes that because it can't see you, then you can't see it. Almost human, you might think.

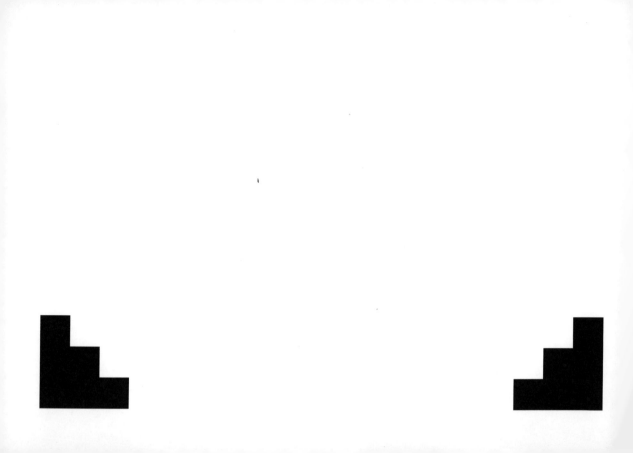

How can Shorty see over Lofty?

The Colosseum is one of the greatest feats of Roman engineering and architecture. It was used for gladiatorial contests, executions, animal hunts, and even mock sea battles. The historian Dio Cassius tells us that 9,000 wild animals were slaughtered at the amphitheatre's inaugural games. As a Roman citizen, short or tall, you wouldn't want to miss such a spectacle.

How do you know, right now, you aren't dreaming?

Freedom belongs to the person who is a little wider awake. And right now that's you.

Why do all roads lead to Baghdad?

Herodotus (c.484 - 425 BC) wrote that asphalt was used in the construction of the towers and walls of Babylon. Early uses of oil abound, yet few could foresee quite how much mankind would come to rely on 'black gold'. Nearly 80% of the world's petroleum still comes from the Middle East, although the world's first commercial oil well was drilled in Poland in 1853.

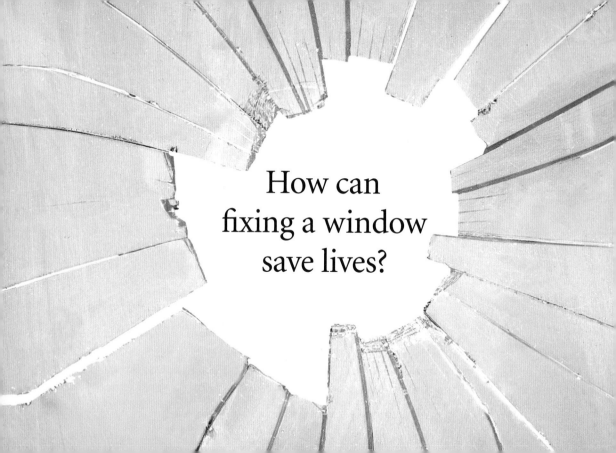

How can
fixing a window
save lives?

In 1982, James Wilson and George Kelling discovered a connection between criminal behaviour and urban environment. Repairing broken windows, cleaning graffiti off walls and clearing the streets of litter reduces robbery, assault and murder rates. The 'Broken Window theory' would be adopted to great affect by Mayor Giuliani in New York after his 1993 election win.

Invent God

How do you make people behave?

1 3 5 7 9 10 8 6 4 2

Published in 2013 by Ebury Press, an imprint of Ebury Publishing. A Random House Group Company. Copyright ©M&C Saatchi Worldwide 2013

M&C Saatchi Worldwide has asserted its right to be identified as the author of this Work in accordance with the Copyright, Designs and Patents Act 1988. All rights reserved. No part of this publication may be reproduced, stored in a retrieval system, or transmitted in any form or by any means, electronic, mechanical, photocopying, recording or otherwise, without the prior permission of the copyright owner.

The Random House Group Limited Reg. No. 954009 Addresses for companies within the Random House Group can be found at www.randomhouse.co.uk

A CIP catalogue record for this book is available from the British Library

The Random House Group Limited supports the Forest Stewardship Council® (FSC®), the leading international forest-certification organisation. Our books carrying the FSC label are printed on FSC®-certified paper. FSC is the only forest-certification scheme supported by the leading environmental organisations, including Greenpeace. Our paper procurement policy can be found at www.randomhouse.co.uk/environment

Printed and bound in Italy by Printer Trento. ISBN 9780091957025

Every effort has been made to contact copyright holders for permissions. Corrections can be made in future printings. Please contact the publishers with any queries.

The publishers wish to thank Kevin Edwards, Alamy, Getty Images and iStockphoto for providing the images.

'Brutal Simplicity of Thought' is a registered trademark of M&C Saatchi